PENGUINS

AF711148

An Animal Information Book

by Elizabeth Elias Kaufman

PRICE/STERN/SLOAN
Publishers, Inc., Los Angeles
1986

Penguins are members of the bird family. Although they have feathers, penguins cannot fly.

There are about seventeen different types of penguins. Some penguins have long tails, and others have crests or feathers that stick out from their heads.

This is a gentoo penguin.

Penguins live along shores or on islands. They spend most of their time in the water. They eat small fish and other small sea life.

All penguins live below the equator. Some live in warm places, but many penguins live in Antarctica where it is very cold.

The largest type of penguin is the emperor penguin. It is more than three feet tall and weighs about eighty pounds.

The smallest type of penguin is the blue penguin, also called the fairy penguin. It is only about fourteen inches tall and weighs about two pounds.

These chinstrap penguins weigh around ten pounds each.

Penguins gather in very large groups before they lay their eggs. These large groups return to the same area every year to lay their eggs. The area they return to is called a rookery. In some cases, there may be a million or more penguins in a single rookery.

Some penguins lay their eggs in nests built out of loose stones. Because there are so many penguins in a rookery, the nests are very close together.

These are Adelie penguins. The mothers and fathers will take turns sitting on the egg until it hatches.

A penguin's eggs do not all hatch at the same time. Sometimes the chicks are hatched a day or two apart. After the chick pecks its way out of the egg, it stays very close to its parents.

A penguin's feathers are waterproof. They protect it from the cold ocean water.

A penguin chick is born with down on its body. Down is a covering of soft, fluffy, baby feathers. These feathers are not waterproof, so a chick cannot go into the water until it loses them and grows adult feathers.

Some types of penguin chicks take up to two weeks to completely lose their down. This is called molting. During that time, the chicks look very strange.

Adult penguins molt their feathers once a year. While they are molting, they cannot go in the water.

Each penguin has a slightly different voice. This helps a penguin find its own mate and its own chick in a rookery containing more than a million penguins.

Penguins are good parents, but they will feed only their own chicks.

Penguins are excellent swimmers. They can dive deep into the water to catch fish.

A penguin's arms are called flippers. Penguins use their flippers to move very quickly through the water.

Penguins can stay under the water for several minutes before coming up to the surface to breathe.

Peruvian penguins are also known as Humboldt penguins. They live along the Pacific coast of South America.

These penguins are just over two feet tall and weigh about ten pounds.

Peruvian penguins are slimmer than penguins that live in colder climates because they don't need as much fat to keep them warm.

The thin black line under its chin gives the chinstrap penguin its name.

Chinstrap penguins often build their nests on rocky cliffs, and use their strong legs and feet to climb up to them.

Gentoo penguins are larger than chinstrap penguins. They are about thirty-two inches tall and can weigh up to fourteen pounds.

Unlike most other penguins, gentoo penguins do not always live in large groups. Gentoos don't build their nests very close together because their rookeries don't contain many penguins.

King penguins do not build nests. Instead, their eggs rest on their feet. The eggs are held against the parent's body to keep them warm. Both parents take turns holding the egg until it hatches.

These Adelie penguins live in Antarctica. They return to their rookeries in October, which is springtime in Antarctica. During this time, the sun shines for almost twenty-four hours every day. During the dark winter, Adelie penguins return to the sea.

Animal Information Books

Titles in this Series

Baby Animals
Baby Zoo Animals
Bears
Big & Little Animals
Birds
Bunnies & Rabbits
Butterflies
Farm Animals
Horses & Ponies
Kittens & Cats
Lions & Tigers
Monkeys & Apes
Penguins
Pets
Puppies & Dogs
Sea Animals
Wild Animals
Zoo Animals

Copyright © 1986 Ottenheimer Publishers, Inc.
Published by Price/Stern/Sloan Publishers, Inc.,
410 North La Cienega Boulevard, Los Angeles, California 90048
All Rights Reserved.
Printed in Brazil.
ISBN: 0-8431-1524-6

*This delightful book will enable your child
stand and relate to the animal world. The
of a young child will easily be satisfie
full-color pictures and easy-to-read type. All
in this series are listed on the last page of this b*

BOOKSRUN
READ.
SELL.
REPEAT.

0
9781644874820
ISBN 0-843-1524-6

SLO
Angeles

$1